M000248937

GOODS OF THE MIND, LLC

Math Challenges

for

Gifted Students

PRACTICE TESTS IN

MATH KANGAROO STYLE

FOR STUDENTS IN GRADES 3-4

Cleo Borac, M. Sc.
Silviu Borac, Ph. D.

This edition published in 2015 in the United States of America.

Editing and proofreading: David Borac, M. Mus.
Technical support: Andrei T. Borac, B.A., PBK

All rights reserved. Except as permitted under the United States Copyright Act, no part of this publication may be reproduced or distributed in any form or by any means, or stored in a database or retrieval system, without prior written permission from the publisher, unless otherwise indicated.

Copyright Goods of the Mind, LLC.

Send all inquiries to:

Goods of the Mind, LLC
1138 Grand Teton Dr.
Pacifica
CA, 94044

Math Challenges for Gifted Students
Level II (Grades 3 and 4)
Practice Tests in Math Kangaroo Style for Students in Grades 3-4

Contents

1 Foreword 5

2 Test Number One 7

3 Test Number Two 15

4 Test Number Three 25

5 Test Number Four 35

6 Test Number Five 45

7 Test Number Six 55

8 Hints and Solutions for Test One 65

9 Hints and Solutions for Test Two 75

10 Hints and Solutions for Test Three   **86**

11 Hints and Solutions for Test Four   **95**

12 Hints and Solutions for Test Five   **101**

13 Hints and Solutions for Test Six   **107**

FOREWORD

This workbook contains six exams that are similar to the Math Kangaroo contest, an international mathematics competition for students in grades 1 through 12. The contents of this book are not associated with the Math Kangaroo organization in any way. None of the problems included has been an actual contest problem. Also, problems are not repeated across the various books in our series. Each workbook presents a completely different set of original problems.

The problems in this book are somewhat more difficult than the ones of the actual competition. In the authors' experience, this is beneficial for preparation. However, the mathematical pre-requisites include only: addition and subtraction with multidigit numbers, multiplication of multi-digit numbers, integer division with remainder, ratios, operations with fractions, percents, rates, and recognition of the basic geometric figures and solids.

As in any contest paper, the difficulty of the items is staggered. The 3-point problems are relatively easy problems based on observation, elementary counting, and reading comprehension. The 4-point and the 5-point problems require more creative applications of the concepts studied in school at the specific grade level.

The authors recommend this book as an additional study material to the series "Competitive Mathematics for Gifted Students" - level 3. As the student progresses through the material of the series, these tests are useful for assessment as well as for training specific competitive skills such as: time management, stamina, and focusing over a longer period of time. We recommend taking one of these tests every month or so. The student should have 75 minutes of contiguous time to solve *without using a calculator*. Using scratch paper is strongly suggested. The student should make diagrams, tables, and show work for each problem.

Copyright Goods of the Mind, LLC, 2014

TEST NUMBER ONE

3-point problems

1. Andy had 28 magnets. He gave all of them, except for the 4 that were larger, to his friend Sandy. How many magnets does he have now?

   (A) 4 **(B)** 8 **(C)** 14 **(D)** 24 **(E)** 32

2. Donkey had 8 straws and Pig had 6 more straws than Donkey. How many straws did they have together?

   (A) 12 **(B)** 14 **(C)** 16 **(D)** 18 **(E)** 22

3. During our vacation at the seaside, we had 5 days with large waves, 5 days of carnival, and 5 days that we spent with new friends that we made on the morning of the second day. At least how many days was our vacation long?

   (A) 5 **(B)** 6 **(C)** 7 **(D)** 15 **(E)** 16

4. Which of the figures labeled from A to E is not a part of the figure with 5 circles?

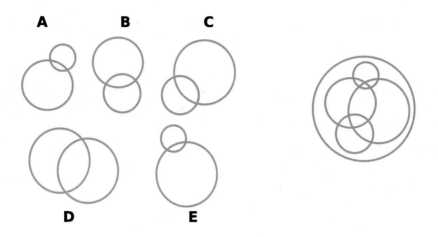

(A) A (B) B (C) C (D) D (E) E

5. When we went on vacation with our friends, they got a room 5 floors above ours. Surprisingly, in the hotel, we met my math teacher, who had a room on the 9th floor, 9 floors below our friends' room. What floor were we on?

(A) 4 (B) 9 (C) 13 (D) 14 (E) 18

6. If 24 is the third of the double of a number, what is that number?

(A) 16 (B) 32 (C) 36 (D) 72 (E) 144

Copyright Goods of the Mind, LLC, 2014

7. ThriftyFlights Airlines has sold tickets for seven sixths of the tickets available for their 294-seat plane. If all the people who purchased tickets show up, how many of them will not find a seat?

(A) 42 **(B)** 49 **(C)** 245 **(D)** 252 **(E)** 343

8. Nick, Tim, and Roger went camping at Bear Lake. Each day, they managed to catch exactly two fish for dinner. Each of them caught at most one fish each day. Nick caught three fish and Roger caught five fish. Nick caught the least number of fish and Roger caught the most fish. At least how many days did they spend camping at Bear Lake?

(A) 5 **(B)** 6 **(C)** 7 **(D)** 8 **(E)** 9

4-point problems

9. Danielle, Andra, and Maggie cleaned out their collections of action figures to put the ones they had extra up for sale. In total, they had 101 figures. Danielle had 5 doubles, Andra 3 doubles, and Maggie had 8 doubles. All their other figures were different. When they set aside the figures they wanted to sell, Danielle was left with twice as many figures as Andra and as many figures as Maggie. How many figures did Andra keep?

(A) 15 **(B)** 17 **(C)** 23 **(D)** 69 **(E)** 85

10. What is the smallest number of popsicle sticks you need in order to make 5 squares of any size?

(A) 4 **(B)** 6 **(C)** 8 **(D)** 16 **(E)** 20

Copyright Goods of the Mind, LLC, 2014

11. Stella's father was 29 when Stella was born, and 33 when Stella's brother was born. Today, their three ages add up to 55 years. How old is Stella?

   (A) 6 (B) 7 (C) 8 (D) 9 (E) 10

12. If some month has three Wednesdays with odd dates, the following month starts on a:

   (A) Saturday (B) Sunday (C) Monday (D) Tuesday (E) Wednesday

13. McDougal's farm and McIvor's farm are 890 yards apart. A canal is built exactly between the two farms, so that they both can use it to water crops. The canal is 40 yards wide and 3 yards deep. At least how many yards of hose must McIvor install in order to be able to pump water from the canal?

   (A) 405 (B) 425 (C) 435 (D) 445 (E) 850

14. At a bus station, only 3 people get on the bus and none get off. At the next stop, 2 passengers get off and none get on. The next stop is the last one, 3 passengers get off and there are no passengers left on the bus. How many people were on the bus before the first stop?

   (A) 2 (B) 3 (C) 4 (D) 5 (E) 8

15. Four towns, Ta, Te, Ti, and To, are connected by a highway that crosses them from East to West, in the given order. The distance between Ta and Ti is 50 miles. The distance between Te and To is 30 miles. The distance between Ta and To is 75 miles. What is the distance between Te and Ti?

   (A) 5 (B) 10 (C) 20 (D) 25 (E) 105

16. Each Rat wears a Hat. When the Rat takes his Hat off, another similar, but smaller Rat, is revealed. When this smaller Rat takes off *his* Hat, another similar, but smaller yet Rat, is revealed. This can keep going on without an end in sight. If 16 Hats were taken off in total, how many Rats can be seen?

    (A) 14 **(B)** 15 **(C)** 16 **(D)** 17 **(E)** 32

5-point problems

17. How many 3-digit positive numbers have consecutive digits in order?

    (A) 7 **(B)** 8 **(C)** 10 **(D)** 14 **(E)** 15

18. Four contestants are participating in a coleslaw preparation contest. Anna needs at least 3 cabbages for the coleslaw recipe, Silvia needs at most 5 cabbages for her recipe, Indira needs no more than 2 cabbages, and Tommy needs fewer cabbages than Silvia. At most how many of their recipes could require the same number of cabbages?

    (A) 0 **(B)** 2 **(C)** 3 **(D)** 4 **(E)** 5

19. Nick computed the sum of all the odd 3-digit numbers that have only one non-zero digit. Andrea computed the sum of all the even 3-digit numbers that have only one non-zero digit. Which of the following statements is true?

    (A) Andrea's sum is larger than Nick's sum by 4500.
    (B) Andrea's sum is larger than Nick's sum by 100.
    (C) Andrea's sum is equal to Nick's sum.
    (D) Nick's sum is larger than Andrea's sum by 100.
    (E) Nick's sum is larger than Andrea's sum by 4500.

Copyright Goods of the Mind, LLC, 2014

20. A tiling has a pattern like the one in the figure:

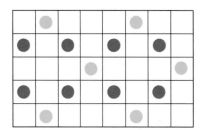

Assuming the pattern continues in the same manner at both ends, without an end in sight, which of the following statements is true?

(A) There are 3 yellow dots for every 4 blue dots.

(B) There is one yellow dot for every blue dot.

(C) There are 2 yellow dots for every blue dot.

(D) There are 3 yellow dots for every 2 blue dots.

(E) There are 5 yellow dots for every 4 blue dots.

21. Harris paid 41 dollars for 8 golden pens and 5 silver pens. Jenna paid 24 dollars for 5 golden pens and 8 silver pens. How much do 2 golden pens and 2 silver pens cost?

(A) 5 **(B)** 9 **(C)** 10 **(D)** 11 **(E)** 13

22. How many even 10-digit positive numbers can Danny write using only 2 ones and 8 zeroes?

(A) 8 **(B)** 9 **(C)** 16 **(D)** 45 **(E)** 90

Copyright Goods of the Mind, LLC, 2014

23. Tommy makes a building by placing cubes on top of each other. The 3×3 grid on the left represents the building as seen from directly above. The numbers represent the number of cubes stacked in each cell of the grid. What does the building look like when viewed from the side in the direction of the arrow?

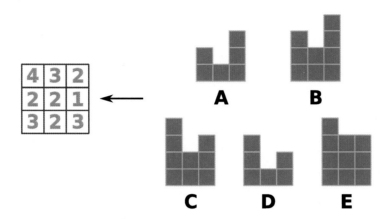

(A) A (B) B (C) C (D) D (E) E

24. A robotic engraver is used to cut square pieces out of a sheet of plastic. Each side of a square counts as one cut. In order to spend less time and energy, the robot is able to figure out how to make the cuts in the shortest possible time. For example, if it has to cut two squares, it will make them adjacent so that one side is shared:

For the example in the figure, the robot will make only 7 cuts. If it had engraved each square separately, it would have made 8 cuts.

An order for 12 squares comes is. How many cuts will the robot make?

(A) 9 (B) 31 (C) 32 (D) 37 (E) 48

Copyright Goods of the Mind, LLC, 2014

Answer Key for Test One

3-point problems	4-point problems	5-point problems
1. A	9. B	17. E
2. E	10. B	18. B
3. B	11. E	19. A
4. D	12. A	20. A
5. C	13. B	21. C
6. C	14. B	22. A
7. B	15. A	23. B
8. B	16. D	24. B

TEST NUMBER TWO

3-point problems

1. How many of the following operations have a non-zero result?

   A. $15 - 5 - 4 - 3 - 2 - 1 =$

   B. $2 \div 2 \times 4 - 12 \div 3 =$

   C. $4 - 2 + 6 - 8 =$

   D. $100 - (100 \div 10) \times 10 =$

   E. $5 \times 5 \times 5 \div 125 - 1 =$

   (A) 0 **(B)** 1 **(C)** 2 **(D)** 3 **(E)** 4

2. In the following correct operation, each of the two stars covers the same number. Which number is it?

   $$19 - \bigstar = 9 + \bigstar$$

   (A) 4 **(B)** 5 **(C)** 6 **(D)** 8 **(E)** 10

3. Daniel has 70 poker chips, Anna has 24 poker chips more than Daniel, and Carla has 33 poker chips less than Anna. How many chips less than Daniel does Carla have?

(A) 9 **(B)** 19 **(C)** 37 **(D)** 57 **(E)** Carla has more chips than Daniel.

4. On the cube, the faces are engraved using three letters: A, B, C, and three digits: 1, 2, 3. Each letter is on a face that is opposite to a face that has a digit on it. Which symbol is on the face that touches the table?

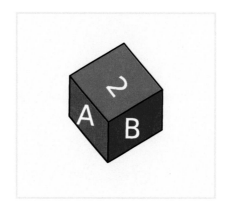

(A) 1 **(B)** C **(C)** 3 **(D)** either 1 or C **(E)** either 3 or C

5. Ilana counts by 5s. The 7th number she says is 40. Which was the 4th number she said?

(A) 10 **(B)** 15 **(C)** 20 **(D)** 25 **(E)** 30

Copyright Goods of the Mind, LLC, 2014

6. How many rectangles are there in the figure?

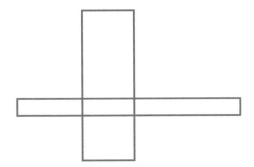

(A) 2 (B) 5 (C) 7 (D) 9 (E) 11

7. Sabrina has only dimes and quarters. What is the difference between the largest and the smallest amounts of coins she can use to pay 85 cents?

(A) 1 (B) 2 (C) 3 (D) 4 (E) 5

8. A regular polygon with 9 sides (*nonagon*) is a polygon with 9 sides of equal length and 9 interior angles of equal measure. Janine has a pair of scissors and wants to cut the polygon into triangles. What is the smallest number of cuts she must make to achieve this?

(A) 5 (B) 6 (C) 7 (D) 8 (E) 9

Copyright Goods of the Mind, LLC, 2014

4-point problems

9. In the following correct multiplications, the digits have been replaced by letters. Each letter corresponds to one digit uniquely. What is the value of Y?

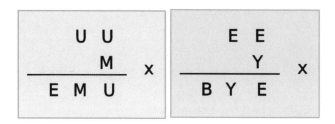

(A) 4 (B) 5 (C) 6 (D) 7 (E) 9

10. Benny, Keira, Lydia, and Rich have the same summer reading list. Rich is reading the book in the middle of the list. Keira is two books ahead of Rich. Lydia is 6 books behind Keira. Benny is 12 books ahead of Lydia. Benny is now reading the last book on the list. How many books are on the list?

(A) 16 (B) 17 (C) 18 (D) 20 (E) 40

11. A pinwheel has five blades. George starts painting the blades using two colors: red and blue. George paints a blade, skips the next blade, and then paints the next blade, and so on. Each time, he switches the color from red to blue. If George started by painting the first blade red, how many more blades will he paint and repaint until he starts painting the same blade in red again for the first time?

(A) 4 (B) 5 (C) 8 (D) 9 (E) 10

Copyright Goods of the Mind, LLC, 2014

12. Delora is in an enrichment group for creative writing. The teacher gave each of the girls 4 pencils and 5 sheets of paper and each of the boys 2 pencils and 4 sheets of paper. After turning in the writing assignment, a quarter of the pencils and a quarter of the sheets of paper remained unused. At least how many students are there in Delora's group?

(A) 6 (B) 8 (C) 15 (D) 28 (E) 63

13. Jim, Jack, and Rich have decided to use a code to message each other. They started by encoding their own names. From the list below, which choices are the ones that represent the codes for their names?

1) ⊗ ∪ ∇

2) ⊗ ◇ ⊏ ∪

3) ⊗ ⊏ ◇ ⊥

4) ⋏ W ◇ $

5) ☾ ∩ ⊏ △

6) ⊗ W ✠

(A) 1, 2, and 3
(B) 1,2, and 4
(C) 1, 3, and 5
(D) 2, 5, and 6
(E) 3, 4, and 6

Copyright Goods of the Mind, LLC, 2014

14. In the following diagram, numbers on higher levels are split into smaller numbers such that there are only whole numbers in all the cells and all the comparison operators are true. What is the number in the square with the question mark?

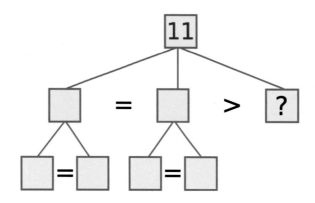

(A) 2 (B) 3 (C) 4 (D) 5 (E) 6

15. Abrianna has 15 blue pens, 12 red pens, and 9 black pens, in her pencil case. If she removes pens from the case without looking, at least how many pens must she remove to be sure she has taken out three pens of the same color?

(A) 3 (B) 6 (C) 7 (D) 10 (E) 22

16. What is the smallest possible result you can get by placing as many parentheses as you like in the following expression?

$$200 - 50 + 40 - 30 + 20 - 10$$

(A) 0 (B) 10 (C) 50 (D) 130 (E) 170

Copyright Goods of the Mind, LLC, 2014

5-point problems

17. Tony, Flora, Tobi, and Fifi are talking about their math homework:

    - 'I would say there are not more than 6 problems to solve,' said Tony.
    - 'Definitely, less than 8,' said Fifi.
    - 'However, not less than 4 problems,' added Tobi.
    - 'We have more than 5 problems to solve,' said Flora.

    How many problems were on their homework?

    (A) 4 **(B)** 5 **(C)** 6 **(D)** 7 **(E)** 8

18. Gina has 20 cards. On each card there is a different number from 1 to 20. What is the largest number of cards Gina should discard if she wants to keep a set of cards with numbers that add up to 100?

    (A) 5 **(B)** 6 **(C)** 7 **(D)** 12 **(E)** 14

19. Tom is 7 and his father is 34. After how many years will Tom be half his father's age?

    (A) 5 **(B)** 10 **(C)** 15 **(D)** 20 **(E)** 22

20. Among all 2-digit positive integers, the sums of the digits of two consecutive numbers differ either by 1 or by which one of the following:

    (A) 0 **(B)** 2 **(C)** 4 **(D)** 6 **(E)** 8

Copyright Goods of the Mind, LLC, 2014

21. Ratberg, the rat, has been placed in a new kind of maze. The maze has many paths made of circles. He can run only on circles. He can change the circle he runs on at any point where two circles touch.

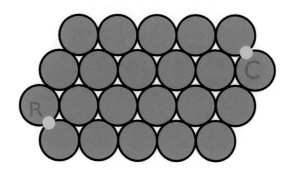

Ratberg is at point R and takes the shortest route to the cheese placed at point C. The path that Ratberg takes is equivalent to how many complete rotations around a single circle?

(A) 1 **(B)** 2 **(C)** 3 **(D)** 4 **(E)** 6

22. An operation consists of taking a positive integer and dividing all its even digits by 2. What is the largest number of such operations that must be performed one after the other on a number before all its digits are odd?

(A) 1

(B) 2

(C) 3

(D) 4

(E) there is no largest number of operations

Copyright Goods of the Mind, LLC, 2014

23. The towns of Hotville, Coolville, and Tepidville are placed on a circular road. There are no other roads that connect them. Based on the distance indicators in the figure, how many miles is the shortest ride between Hotville and Coolville?

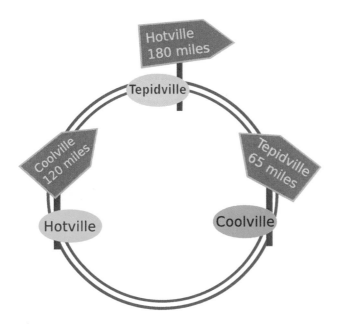

(A) 55 (B) 115 (C) 120 (D) 180 (E) 235

24. A *palindrome* is an integer number that does not change when read backwards. For example: 123321 is a 6-digit palindrome. How many 9-digit palindromes are there that use only the digits 9 and 0?

(A) 8 (B) 9 (C) 16 (D) 18 (E) 81

Copyright Goods of the Mind, LLC, 2014

Answer Key for Test Two

3-point problems	4-point problems	5-point problems
1. A	9. C	17. C
2. B	10. B	18. E
3. A	11. D	19. D
4. B	12. A	20. E
5. D	13. E	21. B
6. E	14. B	22. E
7. C	15. C	23. B
8. B	16. C	24. C

TEST NUMBER THREE

3-point problems

1. What number must be placed in the box to make the equality true?

   $$12 \times 4 = \boxed{} \times 3$$

   (A) 3 **(B)** 4 **(C)** 6 **(D)** 16 **(E)** 24

2. Below each kangaroo there is an expression that shows the number of jumps left before it reaches the finish line. Which kangaroo is in the lead?

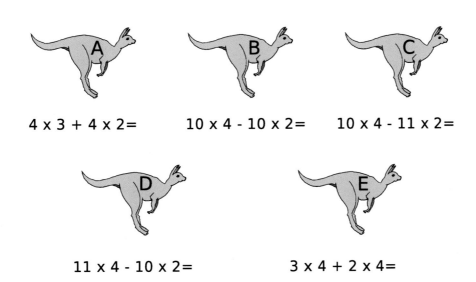

4 x 3 + 4 x 2=   10 x 4 - 10 x 2=   10 x 4 - 11 x 2=

11 x 4 - 10 x 2=   3 x 4 + 2 x 4=

   (A) A **(B)** B **(C)** C **(D)** D **(E)** E

3. Which of the answer choices is a Roman numeral that equals the expression:

$$500 - 100 + 90 - 10 + 5 - 2$$

(A) DCLXXXVII
(B) DCXXCVII
(C) CDLXXXIII
(D) CDXCVII
(E) CDXCVXII

4. If the sequence continues according to a pattern, going from left to right, which number will be on the third yellow balloon?

(A) 18 (B) 21 (C) 24 (D) 27 (E) 30

5. If Danny divides a number by 3 and subtracts 5 from the result, he gets 7. Which number did he start with?

(A) 4 (B) 12 (C) 22 (D) 32 (E) 36

Copyright Goods of the Mind, LLC, 2014

6. In the following sequence of numbers, the sum of any three neighbors is the same. Which number is in the square with the question mark?

9		?			6	12

(A) 6 (B) 9 (C) 12 (D) 18 (E) 27

7. Which of the nets cannot match the cube?

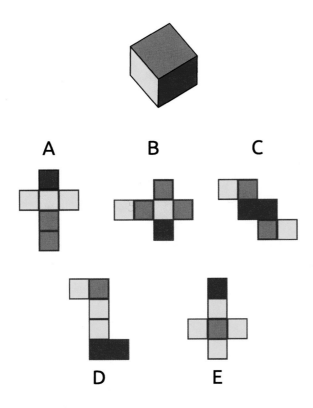

A   B   C

D   E

(A) A (B) B (C) C (D) D (E) E

Copyright Goods of the Mind, LLC, 2014

8. Wanda was 8 years old when Katie was half her age. How old will Katie be when Wanda is 10 years old?

(A) 4 (B) 5 (C) 6 (D) 7 (E) 8

4-point problems

9. In a fruit basket, there are 4 red apples for every 3 yellow apples and 2 yellow apples for every green apple. Which of the following is a possible choice for the number of apples in the basket?

(A) 8 (B) 10 (C) 12 (D) 22 (E) 34

10. Ali was riding his donkey from Marrakesh to Amizmiz. Halfway to Amizmiz, Ali's donkey refused to go forward and Ali decided to return home. They were halfway to Marrakesh when they ran out of water. Ali knew that halfway to Marrakesh, in about 5 miles, there was an oasis and he begged the donkey to get there before the end of the day. About how many miles are there between Marrakesh and Amizmiz?

(A) 5 (B) 10 (C) 15 (D) 20 (E) 40

11. Dylan has a concert ticket for row 24. The row has 80 seats and, looking at it, Daniel notices that all the seats from 1 to 14 are occupied, as well as the seats from 20 to 31, seat 67 and the seats from 70 to 76. How many empty seats can Dylan choose from?

(A) 27 (B) 32 (C) 46 (D) 48 (E) 53

12. In Edith's dance group there are 5 dancers. In how many ways can a group of 3 dancers be selected if Edith must be part of it?

(A) 4 (B) 5 (C) 6 (D) 12 (E) 15

Copyright Goods of the Mind, LLC, 2014

13. Miranda is wrapping presents for the annual prizegiving. She has to put 3 objects in each box. The contents of the boxes do not need to be all the same, but there should not be 2 gifts of the same type in any of the boxes. Miranda has 4 pennants, 4 stuffed animals, 4 transformers, and 3 puzzles. 5 prizes are being awarded. How many of the 5 boxes have the same combination of gifts?

(A) 0 (B) 2 (C) 3 (D) 4 (E) there is no solution

14. In a game, each card has two consecutive numbers on it, one on each face. Amira has three cards and uses them to create a 3 digit number:

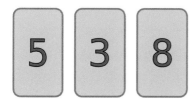

What is the largest odd 3-digit number she can make using these cards?

(A) 385 (B) 967 (C) 963 (D) 964 (E) 976

15. Scott gives candy to his friends for Hallowe'en, but he also receives candy from friends. If Scott gives 5 candies to a friend, he receives 3 candies from that friend. If Scott started with 15 candies, and continued to give out candies for as long as he had at least 5 candies on hand, how many candies did he end up with at the end of the evening?

(A) 0 (B) 1 (C) 2 (D) 3 (E) 4

Copyright Goods of the Mind, LLC, 2014

16. Two numbers can both be divided evenly into groups of 3 as well as into groups of 4. The difference between them can be:

(A) 1 **(B)** 7 **(C)** 18 **(D)** 32 **(E)** 36

5-point problems

17. Michael received four identical bags of marbles. Michael was not able to open one of the bags, but he opened the others. He emptied them and he placed some marbles in one bag, three times as many in another bag, and, in the last bag, he placed five times as many as in the first bag. Which of the following choices could be the number of marbles in the unopened bag?

(A) 4 **(B)** 5 **(C)** 7 **(D)** 11 **(E)** 12

18. Clarissa has a blue paper triangle. She wants to cut it up into 9 smaller triangles, like in the figure. At least how many cuts must she make? (Assume she does not fold the paper or overlap parts of it.)

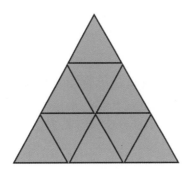

(A) 3 **(B)** 5 **(C)** 6 **(D)** 8 **(E)** 9

Copyright Goods of the Mind, LLC, 2014

19. Kangaroy and Kangarob leave point K at the same time and jump on a line, in opposite directions, at the same speed. In one hour, Kangaroy jumps from K to B and Kangarob from K to C. Kangaroy jumps for 3 hours and Kangarob jumps for 4 hours. Kangaroy rests for 1 hour and Kangarob rests for 3 hours. After resting, each starts jumping towards the other. They will meet for the first time at point:

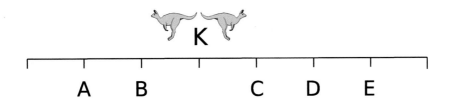

(A) A (B) B (C) C (D) D (E) E

20. 3 mangoes, 4 pineapples, and 2 melons cost the same amount as 7 pineapples. 2 mangoes, 3 melons, and 3 pineapples cost the same amount as 8 melons. How many melons cost the same as 5 mangoes?

(A) 1 (B) 2 (C) 3 (D) 4 (E) 6

21. Two 5-digit numbers have the same sum of digits: 2. The largest possible difference between two such numbers has the sum of digits:
(A) 1 (B) 9 (C) 18 (D) 27 (E) 36

22. Kanga has only 5 dollar bills and Roo has only 10 dollar bills. They try all the different ways in which they could possibly pay 25 dollars. Each time they manage to figure out a new combination of bills that totals 25 they give the money to Dingo, for safekeeping. When they have finished, how many bills does Dingo have?

(A) 3 (B) 7 (C) 10 (D) 12 (E) 15

Copyright Goods of the Mind, LLC, 2014

23. Different even digits are used to make up a 3-digit number and different odd digits are used to make up another 3-digit number. What is the smallest possible difference between the two numbers?

(A) 5 (B) 67 (C) 87 (D) 93 (E) 111

24. Andy and Serena are quizzing each other. For each correct answer, a point is given. For each incorrect answer, a point is deducted. After they answer the same number of questions, the sum of their points cannot be:

(A) 0 (B) 5 (C) 8 (D) 14 (E) 26

Answer Key for Test Three

3-point problems	4-point problems	5-point problems
1. D	9. E	17. E
2. C	10. E	18. D
3. C	11. C	19. D
4. C	12. C	20. C
5. E	13. B	21. D
6. B	14. C	22. D
7. E	15. D	23. A
8. C	16. E	24. B

3-point problems

1. The result of dividing 111111 by 11 is:

   (A) 101 **(B)** 111 **(C)** 1111 **(D)** 10101 **(E)** 10001

2. In the eucalyptus grove there are some families of koalas. Each family consists of parents and two babies. If there are 20 babies in total, how many adults are there?

   (A) 10 **(B)** 20 **(C)** 40 **(D)** 60 **(E)** 80

3. Kangaroy starts on step A and jumps to step E, not skipping any steps. From E, he jumps back to A, and so on, not skipping any steps. After making 21 jumps, what step is Kangaroy on?

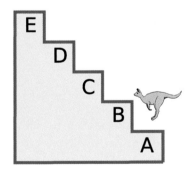

   (A) A **(B)** B **(C)** C **(D)** D **(E)** E

4. Anita has arranged her dolls in alphabetical order by name. On the list, Rapunzel is 15th, Bella is 2nd, Snow White is 20th, Cinderella is 6th, Pocahontas is 13th, Fiona is 9th, and Tiana is 22nd. Which number on the list could Jasmine be?

   (A) 7 **(B)** 8 **(C)** 11 **(D)** 14 **(E)** 16

5. On McIvor's farm there are 10 chickens, 4 dogs, 3 donkeys, and some peafowl. If all these animals total 64 legs, how many peafowl are there?

   (A) 4 **(B)** 8 **(C)** 12 **(D)** 16 **(E)** 20

6. On the map in the figure the transit times between points have been entered in the squares and the waiting times between the different legs of the trip have been entered in the circles. A traveler wants to travel from A to C. Which path has the shortest duration?

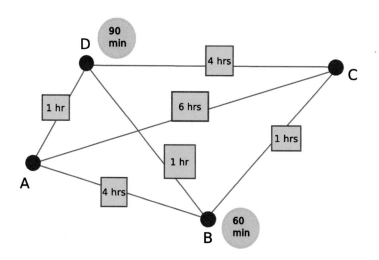

   (A) *ADC* **(B)** *AC* **(C)** *ADBC* **(D)** *ABC* **(E)** *ABDC*

Copyright Goods of the Mind, LLC, 2014

7. Kangarob jumps from square to square from left to right. When he reaches the rightmost square, he jumps back, and so on. Every time he lands on a square he replaces the number written on it by its sum with the number on the square he jumped from. When Kangarob arrives back to the leftmost square, what number does he write on that square?

(A) 13 (B) 14 (C) 16 (D) 28 (E) 29

8. On a full tank of fuel, a cruise ship can travel 450 miles. The "Patagonia" leaves the harbor with a full tank. After a while, only one fifth of the tank of fuel is left. Which of the following is the farthest harbor it can reach without refueling?

(A) Orangeport, 140 miles away

(B) Redport 120 miles away

(C) Greenport, 100 miles away

(D) Tawnyport, 80 miles away

(E) Pinkport, 60 miles away

4-point problems

9. If Dana loses 6 of her magnets, then she will have only 2 more magnets than if she gives half of her magnets to her brother. How many magnets does Dana have?

(A) 0 (B) 4 (C) 8 (D) 16 (E) 20

Copyright Goods of the Mind, LLC, 2014

10. A cube has red and white faces. Each red face is surrounded by white faces. Three such cubes are glued to form a row. Only faces of different colors can be glued together. What is the smallest number of red faces that can be seen? (It is allowed to turn the row of cubes in every way.)

    (A) 0 (B) 1 (C) 2 (D) 4 (E) 6

11. Lars has to swap one of the numbers on the top line and one of the numbers on the bottom line so that the sums of the numbers on each of the two lines are the same. What is the sum of the numbers he must swap?

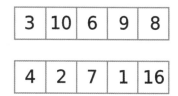

    (A) 7 (B) 10 (C) 12 (D) 17 (E) 24

12. Each of 3 friends has a cat. They meet for tea one day and bring their cats with themselves. When they leave, in how many ways can they select their cats so that each of them leaves with someone else's cat?

    (A) 0 (B) 1 (C) 2 (D) 3 (E) 6

13. Timothy's mother brought home a box of cookies. Timothy went to the kitchen and took 3 cookies. His sister then took half of the remaining cookies. Their friend Katie gave one cookie to Max, and then took another 2 cookies for herself. When Timothy's mother came in the kitchen, she exclaimed: "14 cookies are gone!" How many cookies were there left in the box?

    (A) 2 (B) 4 (C) 5 (D) 6 (E) 9

Copyright Goods of the Mind, LLC, 2014

14. Erik has 3 devices with 4 knobs, 1 device with 3 knobs, and 1 device with 1 knob, as well as a bunch of flexible connectors. Erik uses them to build networks. No device should be connected to itself. All knobs are identical. There is only one connection for each knob. How many different networks are there in the figure?

(A) 0 (B) 2 (C) 3 (D) 4 (E) 5

15. On McIvor's farm there are some chickens. At night, the farmer tries to put them in pens. If he puts 4 chickens in each pen, 7 chickens are left over and there are no more empty pens. If he puts 5 chickens in each pen, there is 1 chicken left by itself in a pen. How many chickens does McIvor have?

(A) 11 (B) 33 (C) 44 (D) 50 (E) 51

16. If we multiply 12 whole numbers, the result is 12. What is the largest possible result if we add the same numbers?

(A) 7 (B) 8 (C) 12 (D) 23 (E) 24

Copyright Goods of the Mind, LLC, 2014

5-point problems

17. In the correct subtraction pictured, what is the largest number that can be subtracted:

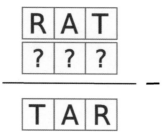

(A) 98 (B) 99 (C) 792 (D) 897 (E) 987

18. Bertie lives in California and his aunt, Mara, lives in Boston. When it is 5 PM in Boston, it is 2 PM in California. If both of them wake up at 7 AM and go to bed at 10 PM, who many hours each day are they both awake?

(A) 8 (B) 9 (C) 10 (D) 11 (E) 12

19. Catrina and Catlina are two playful kittens. They can lap up a quart of milk in 4 days. They invite over Catjim and Catjack, two of their friends. How many days does it take the whole group to finish two quarts of milk? (All kittens have the same appetite.)

(A) 2 (B) 4 (C) 6 (D) 8 (E) 16

Copyright Goods of the Mind, LLC, 2014

20. The seeds of a dandelion have landed as in the figure, where each small square has a side of length 1 unit. However, seeds will not grow into plants if they are too close. To grow into a plant, a seed must be at least 2 units away from all other seeds. In the figure, what is the largest number of seeds that could grow into plants?

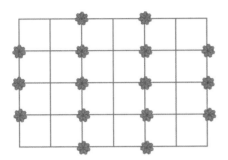

(A) 4 (B) 6 (C) 8 (D) 9 (E) 10

21. Misha's exercise bike has a counter that shows 45678. Misha pedals some more until the counter shows for the first time a number with a digit sum that is five times smaller than the digit sum of 45678. The new counter reading has a digit product of:

(A) 0 (B) 5 (C) 6 (D) 30 (E) it cannot be determined

22. At the big kangaroo and wallaby gathering in Western Australia, kangaroos and wallabies come and go. Every minute 5 of them leave the gathering and 8 of them arrive to the gathering. By how many animals has the gathering increased in half an hour?

(A) 10 (B) 20 (C) 30 (D) 60 (E) 90

Copyright Goods of the Mind, LLC, 2014

23. When number is divided by 11, the remainder is 4 times larger than the quotient. What is the largest such number?

(A) 15 (B) 20 (C) 25 (D) 30 (E) 45

24. Alladin has broken into the den of the Forty Thieves and has found a bundle of keys for the vaults that hide the treasure. Alladin does not know which key opens which vault and he has to use the keys 15 times until he manages to open all the vaults. What is the smallest possible number of vaults in the den?

(A) 3 (B) 5 (C) 6 (D) 10 (E) 15

Copyright Goods of the Mind, LLC, 2014

Answer Key for Test Four

3-point problems	4-point problems	5-point problems
1. D	9. D	17. C
2. B	10. B	18. E
3. D	11. D	19. B
4. C	12. C	20. E
5. B	13. C	21. A
6. C	14. B	22. E
7. D	15. E	23. D
8. D	16. D	24. B

TEST NUMBER FIVE

3-point problems

1. Which number does the star represent?

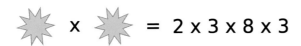 x = 2 x 3 x 8 x 3

   (A) 6 **(B)** 12 **(C)** 16 **(D)** 24 **(E)** 32

2. What is the smallest difference between an odd number and an even number?

   (A) 1 **(B)** 2 **(C)** 3 **(D)** 4 **(E)** different numbers

3. How many 3-digit numbers that end with 7 are greater than 301?

   (A) 54 **(B)** 60 **(C)** 63 **(D)** 70 **(E)** 699

4. Aidan has some tiles that are decorated with circles and he uses them to tile a floor:

Which of the tiles below will produce a tiling with the same pattern?

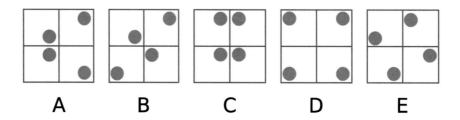

(A) A (B) B (C) C (D) D (E) E

5. Daphne's calculator is broken: it can only perform additions by 4 and multiplications by 3. At least how many operations does Daphne have to make in order to turn 3 into 63?

(A) 2 (B) 3 (C) 4 (D) 12 (E) 21

Copyright Goods of the Mind, LLC, 2014

6. Lisha has a hollow cube made of smaller cubes. How many small cubes are needed to fill it completely?

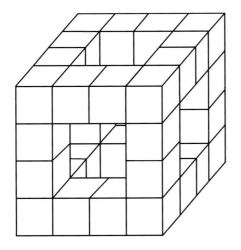

(A) 12 **(B)** 16 **(C)** 24 **(D)** 32 **(E)** 36

7. Erase 5 digits so that the remaining digits form the smallest possible number (the remaining digits should not be moved):

$$10780952043600$$

What is the sum of the 5 digits you erased?

(A) 0 **(B)** 10 **(C)** 15 **(D)** 30 **(E)** 31

8. Kangaroy makes 30 jumps in a minute while Kangarob makes 40 jumps in a minute. Kangarob's jumps are only half as long as Kangaroy's jumps. They both start from under the big eucalyptus when Dingo barks. After 3 minutes, what is the distance between Kangaroy and Kangarob measured in Kangarob's jumps?

(A) 30 **(B)** 40 **(C)** 60 **(D)** 80 **(E)** 120

Copyright Goods of the Mind, LLC, 2014

4-point problems

9. Kangaroy and Kangarob have to find 2-digit numbers so that, when they add the number to the 2-digit number that has the same digits in reverse order, the sum is 88. How many numbers have they found?

(A) 4 (B) 5 (C) 6 (D) 7 (E) 8

10. The perimeter of the small square is 4 and the perimeter of the rectangle is 25. If we cut a small square from each corner of the rectangle, a new figure is formed. What is its perimeter?

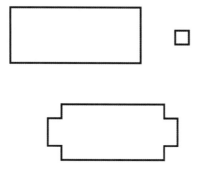

(A) 17 (B) 19 (C) 20 (D) 21 (E) 25

11. The sum of the digits of a number is 10. What is the largest possible product of the digits of this number?

(A) 0 (B) 21 (C) 24 (D) 25 (E) 36

12. Roberto divided a number by its fourth and obtained 4 as a result. What number did he start with?

(A) 1 (B) 4 (C) 8 (D) 16 (E) any number except zero

Copyright Goods of the Mind, LLC, 2014

13. Roberto has 5 cards with symbols on them. He takes a picture of them:

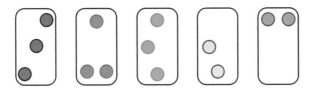

Then, he rotates each card half a circle clockwise and takes another picture:

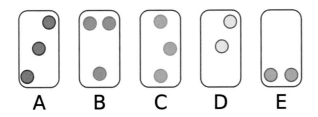

Looking at the picture, Roberto realized his little sister changed one of the cards. Which card did Roberto's sister change?

(A) A **(B)** B **(C)** C **(D)** D **(E)** E

14. A 27-digit number is formed using a pattern of digits:

$$10110111011110111110\cdots 0$$

What is the sum of its digits?

(A) 21 **(B)** 22 **(C)** 25 **(D)** 26 **(E)** 27

Copyright Goods of the Mind, LLC, 2014

15. A jar weighs 250 grams if we fill it with milk and 550 grams if we fill it with glass pellets. If it is only half full with milk, it weighs 150 grams. How much does it weigh if it is half full with glass pellets?

(A) 250 (B) 275 (C) 300 (D) 325 (E) 350

16. What was the date on the first day of the 10th century CE (A.D.)?

(A) 01/01/900
(B) 01/01/901
(C) 01/01/999
(D) 01/01/1000
(E) 01/01/1001

5-point problems

17. Each year, Squirrel Academy enrolls 100 students. Each student spends 3 years learning and obtains a degree at the end of their study. How many students obtained a degree from Squirrel Academy from the year 2006 to the year 2012, inclusive?

(A) 200 (B) 300 (C) 400 (D) 600 (E) 700

18. Dana has a lot of cards with digits on them and she decides to make numbers. She starts at 1 and goes up by one. By the time she gets to 25, she runs out of cards that have 3 on them and decides to replace all the 3s with 7s, of which she has more of. She stops at 100. How many different numbers has she made?

(A) 82 (B) 83 (C) 84 (D) 85 (E) 86

Copyright Goods of the Mind, LLC, 2014

19. More than three numbers are placed in a circle so that any three neighbors are different from one another but have the same sum. What is the smallest number of numbers that can do this?

    (A) 3 **(B)** 4 **(C)** 5 **(D)** 6 **(E)** 7

20. Daniel has some cards with the letters A, B, C, and D on them. With some of his cards, he makes the pattern: ABCDDCBA. Then, he makes up a rule - he reads the pattern from left to right and:

    - if A and B are neighbors, he replaces them both with C
    - if C and D are neighbors, he replaces them both with D

    until the end of the pattern. When he does this one time completely from left to right, he calls this an *operation*. After how many operations does the pattern consist only of Ds?

    (A) 2 **(B)** 3 **(C)** 4 **(D)** 5 **(E)** 6

21. Four strings of different colors are tied at their ends to form a bracelet. How many different bracelets can be made in this way?
    (A) 2 **(B)** 3 **(C)** 4 **(D)** 6 **(E)** 24

22. Kangaroy can hop only in the North, East, South, or West directions. He hops 5 miles East, 6 miles South, and 3 miles West. What is the smallest number of miles he has to hop to get back where he started?

    (A) 2 **(B)** 6 **(C)** 7 **(D)** 8 **(E)** 14

Copyright Goods of the Mind, LLC, 2014

23. In the following cryptarithm, each different digit has been replaced with a different figure. The same figure represents the same digit. Which digit is represented by a triangle?

$$\underline{\hspace{3cm}} \quad 9$$

$$\underline{\hspace{5cm}} \times$$

◀ ◀ ◀ ◀ ◀ ◀ ◀ ◀

(A) 0 (B) 3 (C) 4 (D) 5 (E) 9

24. Wor and Orm, two bookworms, started to tunnel through the two volumes of a book. The two volumes were stacked with their front covers up and with volume one on top of volume two. Wor started from the top and Orm started from the bottom of the stack. They tunnelled at the same speed and met on page 70 of volume two. If the total number of pages is 886, how many pages does volume one have? (Disregard the covers.)

(A) 373 (B) 408 (C) 442 (D) 746 (E) 816

Copyright Goods of the Mind, LLC, 2014

Answer Key for Test Five

3-point problems	4-point problems	5-point problems
1. B	9. D	17. E
2. A	10. E	18. C
3. D	11. E	19. D
4. D	12. E	20. B
5. B	13. D	21. B
6. D	14. A	22. D
7. D	15. C	23. D
8. C	16. B	24. A

TEST NUMBER SIX

3-point problems

1. The smallest multiple of 3 that can be rounded down to 12340 is:

   (A) 12339 **(B)** 12340 **(C)** 12341 **(D)** 12342 **(E)** 12343

2. The number 99 was multiplied by an even number. The result cannot be:

   (A) 0 **(B)** 198 **(C)** 297 **(D)** 396 **(E)** 594

3. The hour hand of a clock rotates 1 full circle around the clock. During the same time, how many complete rotations has the minute hand made?

   (A) 6 **(B)** 10 **(C)** 12 **(D)** 24 **(E)** 60

4. 7 kangaroos run in single file. Once in a while, the last kangaroo moves to the front of the line. How many times must this happen for the kangaroos to be in the same order as at the start?

   (A) 6 **(B)** 7 **(C)** 8 **(D)** 12 **(E)** 14

5. On the number line, the two neighbors of a number have a sum of 136. What is the sum of the digits of that number?

   (A) 6 **(B)** 12 **(C)** 14 **(D)** 67 **(E)** 68

6. Kangarob and Kangaroy jump on the letters of the alphabet. Kangarob starts at A and hops all the way to Z, and Kangaroy hops from Z to A. If they hop at the same time, which letter will Kangaroy be on when Kangarob is on the letter Q?

   (A) F **(B)** G **(C)** H **(D)** I **(E)** J

7. 4 students act in a play that has 6 characters. What is the smallest number of costume changes they might have to do?

   (A) 1 **(B)** 2 **(C)** 4 **(D)** 6 **(E)** 10

8. Some students stand at equal distances from one another in a circle. Ellie and Vijay are standing directly across from one another. Which of the following choices cannot be the total number of students?

   (A) 2 **(B)** 4 **(C)** 6 **(D)** 8 **(E)** 9

   4-point problems

9. Use exactly one time each of the numbers 2, 4, 5, and the operators + and $\times$, as well as any number of parentheses. Which of the following results cannot be obtained?

   (A) 13 **(B)** 14 **(C)** 28 **(D)** 30 **(E)** 38

Copyright Goods of the Mind, LLC, 2014

10. A number of identical 1000 mL empty bottles are placed in a row under identical faucets. The first faucet is turned on. When there are 50 mL in the bottle, the second faucet is started. When there are 50 mL in the second bottle, the third faucet is opened. This process continues until the first bottle is full and then all the faucets are turned off. At this time, which bottle has only 450 mL of liquid?

(A) 9th **(B)** 10th **(C)** 11th **(D)** 12th **(E)** 13th

11. Let's say that a *unit* square is a square formed using 4 neighboring points on a grid:

Consider the number of unit squares we can make using these grid points:

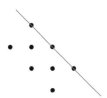

How many more unit squares can we draw if we reflect (mirror) the points across the red line?

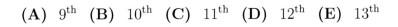

(A) 1 **(B)** 2 **(C)** 3 **(D)** 4 **(E)** 5

Copyright Goods of the Mind, LLC, 2014

12. Two astronomers are looking at a group of 20 stars:

- 'All the stars in this group, except for 9 stars, are red,' said the first astronomer.

- 'All the stars in this group, except for 13 stars, are white,' said the second astronomer.

How many stars are neither white nor red?

(A) 0 (B) 1 (C) 2 (D) 3 (E) 7

13. An equilateral triangle and a rectangle share one side to form a figure. Which of the figures has not been formed in this way?

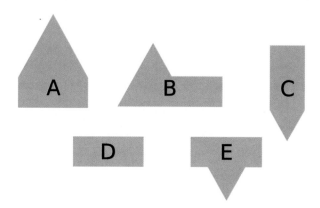

(A) A (B) B (C) C (D) D (E) E

14. Julian has 12 pieces of string. He cuts half of the strings in half. Then, he cuts half of the strings he has in half again. How many pieces of string of any length does he have now?

(A) 18 (B) 24 (C) 27 (D) 33 (E) 36

Copyright Goods of the Mind, LLC, 2014

15. How many of the numbers written using each of the digits 1, 2, 6, and 8 once are multiples of 4?

   (A) 0 **(B)** 2 **(C)** 4 **(D)** 6 **(E)** 8

16. The Borromean Rings are three rings linked in such a way that: if any of the three rings is cut, the other two are no longer linked. Which of the answer choices represents the Borromean Rings?

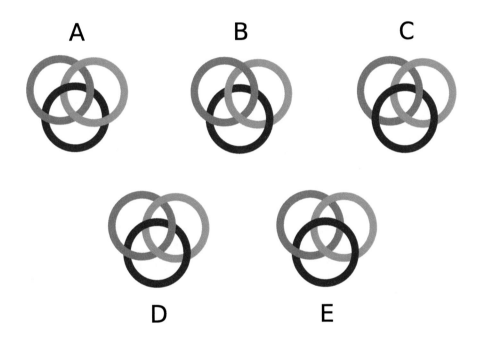

   (A) A **(B)** B **(C)** C **(D)** D **(E)** E

Copyright Goods of the Mind, LLC, 2014

5-point problems

17. Multiples of 3 have been arranged in a pattern:

Row 1	3				
Row 2	6	9			
Row 3	12	15	18		
	21	24	27	32	
	etc.				

In which row will we find the number 111?

(A) 9 **(B)** 10 **(C)** 19 **(D)** 21 **(E)** 35

18. Anne-Lise has 3 cards with a number on each face. For all the cards, the numbers on the two faces add up to the same sum. Five of the numbers are: 4, 6, 8, 9, 10, and the sixth number is unknown. What number from the list is on the same card as the unknown number?

(A) 4 **(B)** 6 **(C)** 8 **(D)** 9 **(E)** 10

19. If we add the triple of a number to five times its double we could obtain:

(A) 8 **(B)** 32 **(C)** 80 **(D)** 91 **(E)** 95

Copyright Goods of the Mind, LLC, 2014

20. The numbers from 100 to 1 have been placed on the vertices of hexagons, following a pattern:

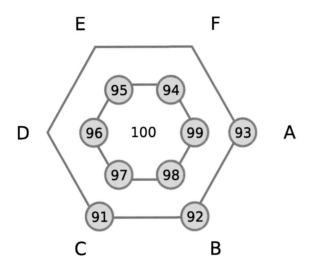

If this pattern continues in the same way, in which direction will the number 1 be placed?

(A) A **(B)** B **(C)** C **(D)** E **(E)** F

21. Kalynka has 4 cards, each with one number on each face. On the back of 1 there is a 2. On the back of 3 there is a 4. On the back of 5 there is a 6. On the back of 7 there is an 8. Kalynka arranges the cards in a pile so that the numbers on the faces that touch add up to the same sum. In how many different ways can she do this?

(A) 0 **(B)** 1 **(C)** 2 **(D)** 3 **(E)** 4

Copyright Goods of the Mind, LLC, 2014

22. Clara Lewin-Gresham gave her best friend a pendant in the shape of a cube with the letters CLGBFF engraved on its faces. From a certain direction, the cube looks like in the picture. Which of the nets is not a possible unfolding of this cube?

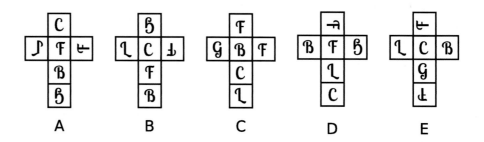

A   B   C   D   E

(A) A (B) B (C) C (D) D (E) E

Copyright Goods of the Mind, LLC, 2014

23. If Nadja would lose 3 of her bracelets, she would have 19 bracelets less than Colleen, who has 3 times more bracelets than Nadja. How many bracelets do Nadja and Colleen have together?

    (A) 24 **(B)** 32 **(C)** 36 **(D)** 48 **(E)** 63

24. A group of students forms a circle. Andra, Bertie, Claude, and Debra are separated by equal numbers of 8 students. Tom, Huck, and Becky are also separated by equal numbers of students. How many students separate Tom from Becky?

    (A) 7 **(B)** 11 **(C)** 12 **(D)** 14 **(E)** 22

Copyright Goods of the Mind, LLC, 2014

Answer Key for Test Six

3-point problems	4-point problems	5-point problems
1. D	9. E	17. A
2. C	10. D	18. D
3. C	11. C	19. D
4. B	12. C	20. C
5. C	13. B	21. C
6. E	14. C	22. E
7. B	15. E	23. B
8. E	16. D	24. B

HINTS AND SOLUTIONS FOR TEST ONE

1. Andy has 4 magnets left.

2. Donkey has 8 straws and Pig has $8 + 6 = 14$ straws. Together, they have $8 + 14 = 22$ straws.

3. If we met our friends on the second day and we spent 5 days with them, then our vacation was at least 6 days long. Since the large waves, the carnival, and the activities with friends can happen at the same time, it is not necessary to have more than 6 days to fulfill all the conditions.

4. Figure D is not a part of the original drawing.

5. Complete the diagram below to figure out the floors on which everyone is located:

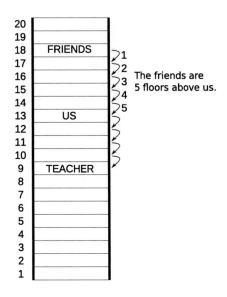

We were on the 13th floor!

6. Work backwards. Multiply 24 by 3 to get the number that 24 is a third of:

$$24 \times 3 = 72$$

Now 72 is the double of the number we are looking for, so we divide 72 by 2 to get 36, which is the correct answer.

7. Since the airline sold one sixth more tickets than the capacity of the plane, one sixth of the capacity is going to equal the number of passengers who cannot find a seat: $294 \div 6 = 49$.

Copyright Goods of the Mind, LLC, 2014

8. Since the fewest and the most fish someone caught are 3 and 5 respectively, Tim must have caught 4 fish. Therefore, the total number of fish caught is 12. Since exactly two fish were caught per day, the vacation at Bear Lake must have lasted 6 days.

9. Danielle will sell 5 figures, Andra will sell 3, and Maggie will sell 8. In total, $5 + 3 + 8 = 16$ figures will be put up for sale. After setting aside the figures for sale, there are $101 - 16 = 85$ figures left. If Danielle has twice as many as Andra and the same number as Maggie, then the figures can be grouped in groups of 5:

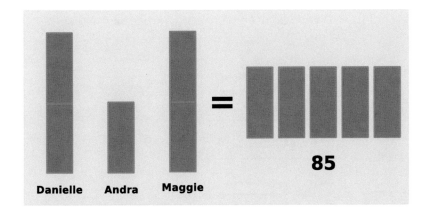

Since $85 \div 5 = 17$, there are 17 figures in each bar. Andra has 17 figures that she wants to keep.

Copyright Goods of the Mind, LLC, 2014

10. Using 6 equal length sticks one can make the following figure, in which 5 squares of different sizes can be found: 4 small squares and one large square.

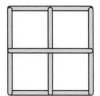

11. Make a timeline to represent the ages of the people involved:

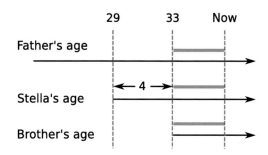

The age difference between Stella and her brother is 4 years. Between the time when the father was 33 and Now the difference of time is represented by the blue segment. Because time flows at the same rate for everyone, all three people have aged the same number of years during the time represented by the blue segment.

From the sum of their ages at the present time, subtract 33 as well as 4 in order to obtain the sum of the three blue segments:

$$55 - 33 - 4 = 22 - 4 = 18$$

This means that the blue segment represents 6 years:

$$18 \div 3 = 6$$

Copyright Goods of the Mind, LLC, 2014

Therefore, at the present time, Stella's brother is 6 years old, Stella is 10 years old, and their father is 39 years old.

12. The only ways a month can have three Wednesdays with odd dates are:

   1. if the month starts on a Wednesday
   2. if the first Wednesday of the month is the 3rd day of the month

In the first case, the month can have 29, 30, or 31 days. The next month can start on: Thursday, Friday, or Saturday.

Su	M	Tu	W	Th	F	Sa
			1	2	3	4
5	6	7	8	9	10	11
12	13	14	15	16	17	18
19	20	21	22	23	24	25
26	27	28	29	30	31	

In the second case, the month must have 31 days:

Su	M	Tu	W	Th	F	Sa
	1	2	3	4	5	6
7	8	9	10	11	12	13
14	15	16	17	18	19	20
21	22	23	24	25	26	27
28	29	30	31			

and the next month starts on a Thursday.
If we consider both cases, the next month cannot start on Sunday, Monday, Tuesday, or Wednesday. Therefore, from the answer choices only Saturday is a possible first day for the next month.

Copyright Goods of the Mind, LLC, 2014

13. Divide the distance between the farms by 2: $890 \div 2 = 445$ yards. From any of the two farms to the midpoint there are at least 445 yards. This is also the midpoint of the canal. The canal extends 20 yards on either side of the midpoint. Subtract 20 yards from 445 to get the shortest distance from either of the two farms to the nearest bank of the canal.

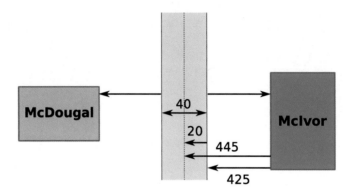

14. Work backwards: put the last 3 passengers back on the bus. Then, put the 2 passengers back. There are now 5 passengers on the bus, but this is after the stop where 3 passengers got on the bus. Before this stop, there were 2 passengers on the bus. Add the bus driver to find that there were 3 people in the bus before the first stop.

15. Make a diagram showing the location of the towns and the given distances:

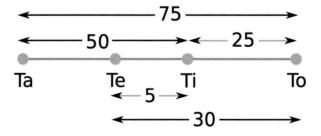

The distances marked in black are the ones given in the statement of

Copyright Goods of the Mind, LLC, 2014

the problem. Use these distances to find the distances marked in blue. First, find the distance from Ti to To:

$$75 - 50 = 25$$

and then use this one to find the distance from Te to Ti:

$$30 - 25 = 5$$

16. Make a diagram similar to the one in the figure, where each Rat is represented by a **C** and each Hat is represented by an **H**. The Hats that have been removed are pictured beside their owners, while the Hat that has not been removed is pictured on top of its owner:

$$
\begin{array}{l}
\overset{\text{H}}{\text{R}} \\
\text{R}\ \text{H} \\
\text{R}\ \text{H} \\
\text{R}\ \text{H}
\end{array}
$$

The last Rat keeps his Hat on because, if he removes it, another Rat will be revealed. Therefore, there are one less removed Hats than Rats can be seen.

In the figure, there are 3 Rats that have removed their Hats and one Rat that still has his Hat on (at the very top). There are 4 cats to be seen in total.

We can solve the same problem for the case with 16 Hats without drawing a diagram anymore. If 16 Hats have been removed, then 17 Rats will be seen. The last Rat will have its Hat on.

Copyright Goods of the Mind, LLC, 2014

17. There are two kinds of numbers with consecutive digits: the ones in which the digits are increasing from left to right, and the ones in which the digits are decreasing from left to right. Because no 3-digit number can start with the digit zero, there are 7 numbers that have 3 consecutive digits in increasing order:

$$123, \quad 234, \quad 345, \quad 456, \quad 567, \quad 678, \quad 789$$

However, since a 3 digit number can have a last digit of zero, there are 8 3-digit numbers with consecutive digits in decreasing order:

$$987, \quad 876, \quad 765, \quad 654, \quad 543, \quad 432, \quad 321, \quad 210$$

There are $7 + 8 = 15$ numbers that satisfy the conditions.

18. Make a chart that shows the numbers of cabbages each could use:

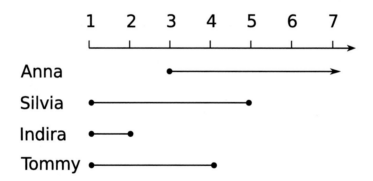

Silvia and Tommy cannot have the same number. Indira and Anna cannot have the same number. This leaves only possibilities with different pairs of competitors that could have the same number: Tommy and Anna, Silvia and Anna, Tommy and Indira, Silvia and Indira.
At most two of the recipes could have called for the same number of cabbages.

19. All the three digit numbers that have only one non-zero digit end in zero and are, therefore, even. Nick's sum is equal to zero. Andrea's

sum is:

$$100 + 200 + 300 + 400 + 500 + 600 + 700 + 800 + 900 = 4500$$

The correct answer is (A).

20. Isolate a fragment of the pattern that repeats as you build the pattern from left to right (or right to left). An example is provided in the figure, but there are other correct choices:

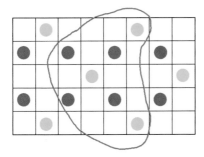

Overall, there are 3 yellow dots to every 4 blue dots. The correct answer is (A).

21. Together, Harris and Jenna purchased 13 golden pens and 13 silver pens, paying a total of $41 + 24 = 65$ dollars. Since $65 = 5 \times 13$, one gold pen and one silver pen cost together 5 dollars. Two gold pens and two silver pens cost 10 dollars.

22. For the number to be a 10-digit number, the first (leftmost) digit must be equal to 1. Since the number must be even, the last (rightmost) digit must be equal to 0. There are 8 spaces left for Danny to place the remaining 1. The remaining spaces will be automatically filled with zeroes. Therefore, Danny can write 8 different numbers.

23. In each row, the tallest tower will show even if it is behind other, smaller, towers. Make sure to look from the direction of the arrow. The correct answer is (B).

Copyright Goods of the Mind, LLC, 2014

24. The robot will place the squares so they share as many shared sides as possible. This happens if it places the squares in a 3 by 4 block (or 4 by 3 - the rotated figure is equally correct):

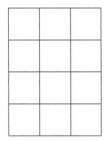

In this case, it will make $3 \times 5 + 4 \times 4 = 15 + 16 = 31$ cuts.

The other options for combining the squares are not as good:

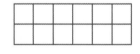

The array of 2 by 6 squares needs $3 \times 6 + 7 \times 2 = 18 + 14 = 32$ cuts, while the array of 1 by 12 needs $2 \times 12 + 1 \times 13 = 24 + 13 = 37$ cuts.

If 12 squares are engraved individually, the number of cuts is equal to $4 \times 12 = 48$.

Copyright Goods of the Mind, LLC, 2014

1. All the operations result in zero.

2. The star covers the number 5:

$$19 - 5 = 9 + 5$$

3. Anna has 94 chips, Carla has $94 - 33 = 61$ chips. Carla has 9 chips less than Daniel.

4. Since there is a 2 on the top face, the bottom face must have a letter on it. Since A and B are already visible on some lateral faces, the only possibility left for a letter is a C.

5. Make a table with the numbers Ilana said, and each of their counts. It will be easier if you list them backwards:

40	35	30	25	20
7th	6th	5th	4th	3rd

6. Count rectangles of all sizes: there are 5 rectangles that are not intersected by any other rectangles. Also, there are 2 large rectangles, and 4 rectangles that are formed by two smaller rectangles each:

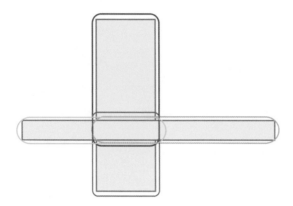

There are total of $5 + 2 + 4 = 11$ rectangles of all sizes.

7. The largest amount of coins she use to pay 85 cents, is 7 - 6 dimes and one quarter:

$$10 + 10 + 10 + 10 + 10 + 10 + 25 = 85$$

The smallest amount of coins is 4 - 3 quarters and one dime:

$$25 + 25 + 25 + 10 = 85$$

The difference is: $7 - 4 = 3$ coins.

8. Any convex polygon with N sides can be cut into at least $N - 2$ triangles. For this, $N - 3$ cuts are necessary. In the figure, a nonagon is represented, together with the 6 cuts that dissect it completely into triangles (*triangulates it*):

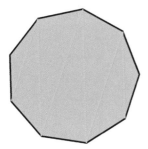

9. The two multiplications look similar, so let us do the reasoning for the left hand one and see what happens. Since the last digit occurs unchanged in the product, there are only two possibilities:

   - M is 1, *or*
   - $U = 5$ and $M = 9$ *or*
   - $U = 4$ and $M = 6$.

   If $M = 0$ the resulting number cannot be a 3-digit number, it would simply be UU.

   Therefore, if $U = 5$ and $M = 9$, the multiplication is $55 \times 9 = 495$. If $U = 4$ and $M = 6$, the multiplication is $44 \times 6 = 264$. This second multiplication would correspond to the right hand side cryptarithm, since the letter E is shared between the two. Therefore, $E = 4$ and $Y = 6$.

10. If Benny is 12 books ahead of Lydia, then he is 6 books ahead of Keira, and 8 books ahead of Rich. If there are 8 books on the list after the book that Rich is currently reading, then there are another 8 books on the list before the book that Rich is reading, because we know that Rich is now reading the book that is in the middle of the list. There are 17 books in total: 8 before Rich's book, 8 after, and the one he is reading now.

Copyright Goods of the Mind, LLC, 2014

11. Make a diagram of the pinwheel. Notice that, because the number of blades is odd, the same blade will be painted a different color the next time around.

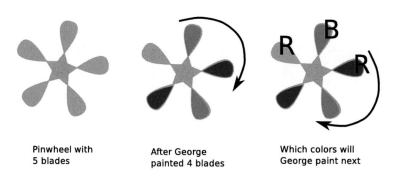

Pinwheel with
5 blades

After George
painted 4 blades

Which colors will
George paint next

George will paint the starting blade in red again after two complete turns. By that time, he has painted all other blades twice (8 painted blades) and he repainted the starting blade only one time, in blue, to a total of 9 blades. We did not count the initial blade because the question asks for *how many more* blades will be painted. We also did not count the time when George painted the first blade red again because the question says *until*.

12. The total number of pencils must be a multiple of 4, otherwise a quarter of the pencils cannot be represented by an integer number. Likewise, the number of sheets of paper must be a multiple of 4. Notice that the remaining supplies are counted 'after turning in the assignment,' and may have been returned by either boys or girls. However, it is necessary for the total number of students to be a multiple of 3, since three quarters of each supply were used.

Since there are 4 pencils for each girl, and 2 pencils for each boy, the number of boys must be a multiple of 2. The least number of boys is 2, from the perspective of the number of pencils.

Copyright Goods of the Mind, LLC, 2014

Since there are 4 sheets of paper for each boy, the number of sheets given out to boys is a multiple of 4. Not so with the number of sheets given out to girls, which is five times the number of girls. The number of girls must be a multiple of 4. The least number of girls is 4.

The least number of students is $2 + 4 = 6$. 6 is a multiple of 3, therefore we can stop our search for a solution here.

13. We have to look for items that have identical characters in the same locations as in the names that are encoded:

This is only satisfied by the combination:

6) ⊗ W ✠

3) ⊗ C ◇ ⊥

4) 人 W ◇ $

Copyright Goods of the Mind, LLC, 2014

14. To split a number into two equal numbers, it has to be even. Therefore, 11 must be split into two equal even numbers and the mystery number. This is possible in only two ways:

$$11 = 4 + 4 + 3$$
$$11 = 2 + 2 + 7$$

Of these, only the first one satisfies the inequality. The solution is:

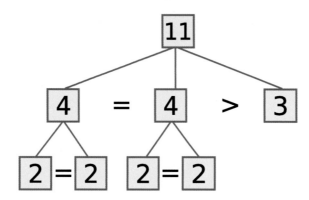

15. Apply the Pigeonhole principle: there are 3 boxes (the three colors) and more than enough pigeons. To be sure that there is at least one color she has 3 pens of, she has to remove at least 7 pens: in the most unfavorable case, she will get 2 pens in each box for a total of 6 pens. In this case, any new pen is going to be of some color, so that there are 3 pens of the same color.

16. We get the smallest result by subtracting as large a number as possible:

$$200 - (50 + 40) - (30 + 20) - 10 = 200 - 90 - 50 - 10 = 50$$

Copyright Goods of the Mind, LLC, 2014

17. Make a diagram to help you understand the exact meaning of the statements:

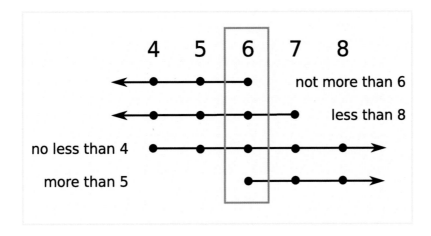

The only number that satisfies all the conditions is 6.

18. Gina must keep the smallest possible number of cards. By adding together the largest numbers:

$$20 + 19 + 18 + 17 + 16 = 90$$

we notice that it is possible to obtain 100 immediately by selecting 10 to be part of the set. Therefore, Gina needs at least 6 cards to have a set of numbers that totals 100. She must discard 14 cards.

Copyright Goods of the Mind, LLC, 2014

19. Tom and his father age at the same time. After an unknown number of years, Tom will be half as old as his father. Make a diagram. In our diagram, the blue box is the number of years that passed. Before then, Tom was 7 and his father was 34.

Add a number of years to Tom's age.

7	

Multiply Tom's new age by 2.

7		7	

This is equal to adding the same number of years to the father's age.

34	

Remove one blue box from both boxes that are equal:

Compare the boxes to find that the blue box must be of size 20.

20. The sums of the digits of two consecutive numbers generally differ by 1, unless there is a change in the tens digit, in which case the sums differ by 8. The units digit resets to 0 from a value of 9, which decreases the sum of the digits by 9, and the tens digits increases by 1, which increases the sum of the digits by 1. Examples:

$$19, \ 20 \ \longrightarrow \ S(19) = 10, S(20) = 2$$
$$29, \ 30 \ \longrightarrow \ S(29) = 11, S(30) = 3$$

21. Notice that each circle touches neighboring circles at 6 points. This means that, by running from a point where two circles touch to the next closest point where two circles touch, Ratberg travels the sixth part of a whole circle.

In the figure, one of the possible shortest paths from R to C is sketched using purple dots. Another such path is sketched using blue dots. Therefore, there are several different paths that are shortest. For any of them, Ratberg has to run 12 times the arc between two closest dots - that is the same as running twice around one of the circles.

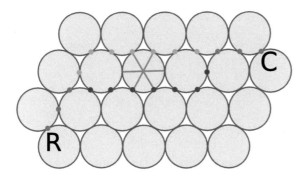

22. The even digits are: 0, 2, 4, 6, and 8. Since the result of dividing 0 by any number is 0, it is not always possible to turn all the digits into odd digits. Should the number in question contain the digit 0, there is no number of operations that can turn it into a number with all odd digits.

23. In the direction of the arrows, the distance between Hotville and Coolville is 120 miles and the distance from Tepidville to Hotville is 180 miles. If we add these two routes, we cover the entire circle once, plus an extra stretch of road between Tepidville and Coolville. Between Tepidville and Coolville there are 65 miles, which is the distance that was covered twice. Therefore, get the total length of the circular road (circumference) by executing the operations:

$$180 + 120 - 65 = 300 - 65 = 235 \text{ miles}$$

If we subtract the distance between Hotville and Coolville in the direction of the arrow from the total circumference, we obtain the distance between Hotville and Coolville in the other direction around the circle:

$$235 - 120 = 115 \text{ miles}$$

Compare the two distances: $115 < 120$!

24. The first digit must be 9 - therefore, the last digit must also be 9. There are two choices for the second digit, but the choice for the digit before last must be the same. There are 2 choices for the third digit, 2 choices for the fourth, and 2 choices for the fifth. The remaining digits are already known now. Make a diagram with the number of possible choices for each digit:

Copyright Goods of the Mind, LLC, 2014

Number of choices we have for each place value:

1 2 2 2 2 1 1 1 1

Examples of numbers:

9 0 0 9 0 9 0 0 9

9 9 9 0 9 0 9 9 9

The total number of ways we can choose the digits is $2 \times 2 \times 2 \times 2 = 16$. This is also the total number of possible palindromes of this type.

Copyright Goods of the Mind, LLC, 2014

1. Divide 12 by 3 and get that $4 \times 4 = 16$.

2. Notice rapidly that choices A and B total to the same result. Since, in choice C, the number subtracted is larger than the number subtracted in B - from the same quantity - then C must be lower than A and than B. In D, the reverse is true, the number to subtract from is increased - therefore, this result must be larger than A, B, and C. D is the same as A.

   In conclusion, the kangaroo that has the smallest distance left until it reaches the finish line is C.

3. $500 - 100 = 400$ which is represented as CD.

   $90 - 10 = 80$ which is represented as LXXX.

   $5 - 2 = 3$ which is represented as III. The answer is CDLXXXIII.

4. The distance between two yellow balloons is $15 - 6 = 9$. Therefore, the number on the next yellow balloon should be $15 + 9 = 24$.

5. Work backwards:

   Reverse the subtraction by adding 5 to the result: $7 + 5 = 12$.
   Reverse the division by multiplying the result by 3: $12 \times 3 = 36$.

6. Represent the number to the left of the 6 by a circle. The circle, the 6, and the 12 must add up to the number that is the sum of any three neighboring numbers. Therefore, the number on the left of the circle must be 12. Propagate this scheme further to the left, to find out that the number represented by the circle must be 9:

| 9 | | | ? | | | | 6 | 12 |

| 9 | | | | | | ● | 6 | 12 |

| 9 | | | | 12 | ● | 6 | 12 |

| 9 | | 12 | ● | 6 | 12 | ● | 6 | 12 |

7. In any of the nets, there must be a corner surrounded by faces of different colors. This is true about all the nets except for net E where there are 2 yellow faces on each corner. The colors of the other faces are not relevant, since we cannot see them.

8. Katie was 4 when Wanda was 8. Wanda will be 10 in two years, and Katie will be 6.

9. If the yellow apples can be grouped in groups of 3 as well as in groups of 2, then the number of yellow apples must be a multiple of 6. We can infer the number of red and green apples from the number of yellow apples. Make a table with the first few possible solutions:

Yellow	Red	Green	Total
6	8	3	17
12	16	6	34

Copyright Goods of the Mind, LLC, 2014

The table need not be any longer (although there are, of course, an infinite number of solutions), because we can already spot one of the answer choices. All other solutions are larger than 34.

10. Make a simple diagram to illustrate Ali's adventures:

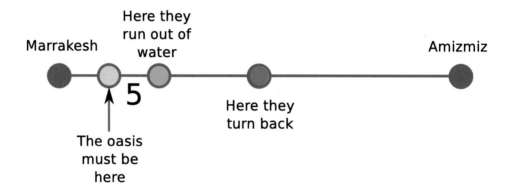

From Ali to the oasis there are 5 miles and, from the oasis to Marrakesh, there are another 5 miles. From Marrakesh to the point where they turn back there are $10 + 10 = 20$ miles. From Marrakesh to Amizmiz there are $20 + 20 = 40$ miles.

11. There are:

   14 seats with numbers from 1 to 14

   $31 - 20 + 1 = 12$ seats with numbers from 20 to 31

   1 seat with the number 67

   7 seats with numbers from 70 to 76.

The total number of occupied seats is: $14 + 12 + 1 + 7 = 34$ seats. Of the 80 seats, $80 - 34 = 46$ are empty.

Copyright Goods of the Mind, LLC, 2014

12. If Edith is part of the group, then only two other dancers must be selected. If the dancers are A, B, C, D, we can select two dancers in 6 different ways:

    AB, AC, AD

    BC, BD

    CD

13. There are 4 types of objects and each box has 3 different objects in it. Since there are only 3 puzzles, there must be two boxes without puzzles that will, therefore, have identical contents (pennant, animal, transformer). However, we have to make sure that there is a solution. Such a solution exists:

    animal, transformer, puzzle

    pennant, puzzle, animal

    pennant, puzzle, transformer

14. The numbers on the other faces of the cards can be: 7 or 9 for the card that shows 8, 2 and 4 for the card that shows 3, and 4 and 6 for the card that shows 5. The largest odd number she can make is 963.

15. Make a table to summarize the operations with candies:

    - Scott gives the 15 candies to 3 of his friends and got 9 candies from them.
    - Scott gives 5 candies to a friend and receives 3 candies. He now has 7 candies.
    - Scott gives 5 candies and receives 3 candies. He now has 5 candies.
    - Scott gives 5 candies and receives 3 candies.

    Scott now has less than 5 candies. Since he started with an odd number, at each step he will have the next smaller odd number. The sequence stops when he has less than 5 and the smaller odd number after 5 is 3.
    Alternate solution: Each time Scott gives and receives candies, the total number of his candies decreases by 2.

Copyright Goods of the Mind, LLC, 2014

16. If the numbers are multiples of 3 as well as of 4, then they are both multiples of 12. The difference between them is a multiple of 12. The only multiple of 12 among the answer choices is 36.

17. Represent the number of marbles Michael put in the first bag by a box. Then, the second and the third bag contain 3 and 5 boxes of the same size, respectively:

Move two of the boxes so that there is an equal number of marbles in each bag:

Therefore, the number of marbles in each bag must be a multiple of 3. The only multiple of three in the choices is (E).

Copyright Goods of the Mind, LLC, 2014

18. She must make at least 8 cuts:

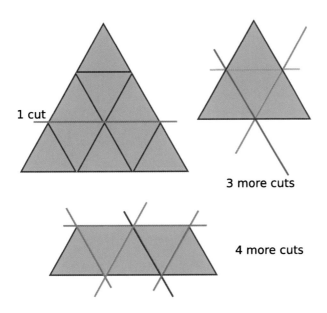

Notice that, once a cut has been made, the two resulting pieces of paper are detached from one another.

19. Kangaroy will reach his resting point when Kangarob is at point E. Kangaroy will rest until Kangarob reaches his resting point. Kangaroy will reach point K when Kangarob will start jumping back. At this time, they are separated by four hours of jumping. They will meet at D.

Copyright Goods of the Mind, LLC, 2014

20. From the data, one can infer that 3 mangoes and 2 melons cost the same as 3 pineapples.

Also, 2 mangoes and 3 pineapples cost the same as 5 melons.

From here, use a diagram:

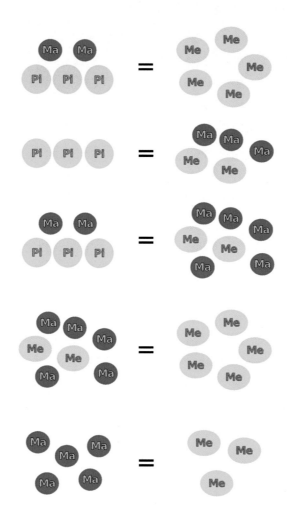

21. The largest possible difference happens when we subtract the smallest possible 5-digit number that has a digit sum of 2 (10001) from the largest possible 5-digit number that has a digit sum of 2 (11000.)

$$11000 - 10001 = 999$$

The digit sum of the difference is: $9 + 9 + 9 = 27$.

22. The possibilities are:

$$
\begin{aligned}
5 \times 5 &= 25 \\
5 \times 3 + 1 \times 10 &= 25 \\
5 \times 1 + 2 \times 10 &= 25
\end{aligned}
$$

They have given Dingo $5 + 3 + 1 = 9$ bills of 5 dollars and $1 + 2 = 3$ bills of 10 dollars. Dingo now has a total of $9 + 3 = 12$ bills.

23. To obtain the smallest possible difference we have to have the smallest possible difference between the hundreds digits such as, for example: $9 - 8$, $6 - 5$, etc. On the other hand, the difference between the tens digits has to be the largest in order to borrow as much as possible. The difference between units digits must also be the largest possible, for the same reason. However, the digits cannot repeat and we can only maximize with this constraint. The only solutions for the smallest difference of 5 are:

$$
\begin{aligned}
602 - 597 &= 5 \\
402 - 397 &= 5
\end{aligned}
$$

Copyright Goods of the Mind, LLC, 2014

24. Partition the quizzing into rounds. One round consists in one question for Andy and one question for Serena. They both start with zero points. After one round the total number of points can be one of the following:

$$
\begin{aligned}
(-1) + (-1) &= -2 \\
1 + 1 &= 2 \\
1 - 1 &= 0
\end{aligned}
$$

Therefore, after one round the total number of points remains *even*: from 0 it changes to one of $\{-2, 0, 2\}$. After any subsequent round, the total number of points can only decrease by 2, increase by 2, or remain the same. This means that the total number of points is always even, no matter how many questions there are, or how they answer.

Copyright Goods of the Mind, LLC, 2014

1. $10101 \times 11 = 111111$

2. The number of adults is equal to the number of babies, because there are two parents for every two babies.

3. Kangaroy needs 4 jumps to go from A to E and 4 jumps to go from E to A. Since $20 = 4 * 5$, Kangaroy will be on step D after 21 jumps, going downward.

4. Place the names in alphabetical order. J must be between F(9) and P(13). The only answer choice that satisfies this condition is (C).

5. 10 chickens have 20 legs. 4 dogs have 16 legs. 3 donkeys have 12 legs. Add all these legs: $20 + 16 + 12 = 48$, and subtract from the total: $64 - 48 = 16$. Each peacock or peahen has 2 legs. Therefore, there are $16 \div 2 = 8$ peafowl on the farm.

6. Calculate the total travel time for each path, in hours:

ADC	$1 + 1.5 + 4 = 6.5$
AC	6
ABC	$4 + 1 + 1 = 6$
ADBC	$1 + 1.5 + 1 + 1 + 1 = 5.5$
ABDC	$4 + 1 + 1.5 + 4 = 10.5$

   The shortest time from A to C is achieved by following the path $ADBC$.

7. Follow Kangarob's jumps and perform the additions:

1	1	1	1	1	1	1

1	2	3	4	5	6	7

28	27	25	22	18	13	7

8. $450 \div 5 = 90$. The "Patagonia" can reach as far as Tawnyport.

9. Make a diagram that models the word problem:

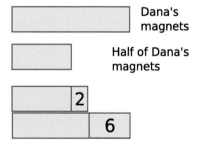

From the diagram, we notice that Dana has 16 magnets, because $6+2 = 8$ is equal to half of the magnets.

Copyright Goods of the Mind, LLC, 2014

10. There are a minimum of one red face and a maximum of two red faces on each cube. If we want the smallest number of red faces, we will choose to have only one red face on each cube. Also, exactly 2 red faces will be glued to exactly 2 white faces. There will be 3 red faces in total and 2 sections that we glue. Since only faces with different colors can be glued, 2 of the red faces will be glued and only 1 will be left to face outward.

11. The sum of the first row is: $3 + 10 + 6 + 9 + 8 = 36$. The sum of the second row is: $4 + 2 + 7 + 1 + 16 = 30$. 3 units must be transferred from the first row to the second to make the sums equal. The only pair of numbers that differ by 3 are 10 and 7.

12. Denote the friends with A, B, C, and the cats with 1, 2, 3. Cat 1 belongs to A, cat 2 belongs to B, and cat 3 belongs to C:

A	B	C
1	2	3
3	1	2
2	3	1

There are only two possibilities for cat 1: B and C. For each of these, there is only one way we can assign the other two cats.

13. Since 14 cookies are gone, $14 - 3 = 11$ were taken by Timothy and his sister. Since Timothy took 3, his sister must have taken $11 - 3 = 8$ cookies. This means there were $8 \times 2 = 16$ cookies in the box after Timothy took 3 cookies. Therefore, there were $16 + 3 = 19$ cookies in the box at the start. If 14 cookies are gone, then there must be 5 cookies left in the box.

Copyright Goods of the Mind, LLC, 2014

14. Only two configurations are possible: the one in which the device with 1 knob is connected to the device with 3 knobs, and the one in which the device with 1 knob is connected to a device with 4 knobs.

15. From the 7 chickens left over, place one more chicken in each pen, to make up to 5 chickens per pen. Then, remove 3 chickens from the last pen and place each one in some pen that still has only 4 chickens in it, continuing to make 5 chickens per pen. Now all pens have 5 chickens in them and there is one chicken left over by itself in the last pen. Therefore, there are 10 full pens and 1 pen with a single chicken in it. The total number of chickens is $5 \times 10 + 1 = 51$.

16. The largest possible sum can be obtained if all the numbers are 1 except for a single number, which is 12:

$$1 \times 1 \times 1 \times 1 \times 1 \times 1 \times 1 \times 1 \times 1 \times 1 \times 1 \times 12 = 12$$
$$1 + 1 + 1 + 1 + 1 + 1 + 1 + 1 + 1 + 1 + 1 + 12 = 23$$

17. The digit represented by **T** must be as small as possible. However, it cannot be zero, since it is also the leftmost digit of the result. We will assign $T = 1$.

The digit represented by **R** must be as large as possible, therefore $R = 9$.

It is easy to verify that, no matter which digit we use in the place of A, the difference is the same. For example,

$$901 - 109 = 792$$

Copyright Goods of the Mind, LLC, 2014

18. Make the following diagram of the special times in Bertie's and Mara's day:

Mara	Bertie
7 AM	4 AM
10 AM	7 AM
10 PM	7 PM
1 AM	10 PM

Between Bertie's wake-up time and the time Mara goes to bed, there are 12 hours during which they are both awake.

19. We double the number of kittens, the milk will be consumed twice as fast. 4 kittens will drink one quart in 2 days, and 2 quarts in 4 days.

20. Select the seeds as in the figure:

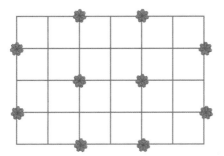

21. The digit sum of 45678 is 30. A digit sum that is 5 times smaller is equal to 6. The next number that has a digit sum of 6 is 50001. It has

Copyright Goods of the Mind, LLC, 2014

a digit product of 0.

22. Each minute, the number of animals increases by 3. After 30 minutes, the number of animals increases by $3 \times 30 = 90$.

23. The remainder cannot be larger than 10. Therefore, the quotient cannot be larger than 2. The largest number is found by assuming the largest possible quotient:

$$11 \times 2 + 8 = 22 + 8 = 30$$

24. To find the smallest possible number of vaults we have to assume the largest number of unsuccessful attempts to open. Assuming there are 5 vaults, Alladin may have to try all the keys until he opens the first one. After opening the first one, he may have to try each of the remaining 4 keys until he opens the second vault. And so on, until the total number of times Alladin uses a key becomes:

$$5 + 4 + 3 + 2 + 1 = 15$$

Note that the initial assumption that there may be 5 vaults has been inspired by the observation that 15 is a triangular number, i.e. it can be obtained by adding consecutive whole numbers starting from 1.

Copyright Goods of the Mind, LLC, 2014

HINTS AND SOLUTIONS FOR TEST FIVE

1. Re-order the factors so you have two identical groups:

$$2 \times 3 \times 8 \times 3 = 4 \times 3 \times 4 \times 3$$

2. Consecutive numbers differ by 1. Since the odd and even numbers alternate, the closest distance between an odd and an even number is 1. Notice that the concepts of even and odd apply only to integer numbers.

3. 70 numbers. There are 7 choices for the hundreds digit: 3, 4, 5, 6, 7, 8, and 9. There are 10 choices for the tens digit: any of the ten digits can be used in this place. There is one single choice for the ones digit: 7. The total number of choices is $7 \times 10 = 70$.

4. The tiling with the choice D is the same as the one with the original tile model:

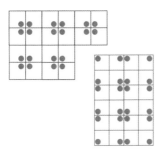

5. Daphne will add $3 + 4 = 7$ and then multiply $7 \times 3 = 21$, followed by the multiplication $21 \times 3 = 63$. Other sequences of operations are possible, but they consist of more operations:

$$3, 7, 21, 63$$
$$3, 9, 13, 17, 21, 63$$
$$3, 9, 13, 17, 51, 55, 59, 63$$
$$3, 9, 27, 31, 35, 39, 43, 47, 51, 55, 59, 63$$
$$3, 7, 11, 15, 19, 23, 27, 31, 35, 39, 43, 47, 51, 55, 59, 63$$

6. Organize your counting so that it is easy and reliable. For example, count the corners first: there are 8 corners and there is one small cube in each. Then, figure out that there are 2 cubes needed to connect the corners along the cube's edges. Since there are 12 edges, there are 24 small cubes that connect the corner cubes. In total, there are $24 + 8 = 32$ cubes that make up the hollow cube.

A full cube of the same size has 4 layers of 16 cubes each, in total 64 cubes. Lisha needs $64 - 32 = 32$ small cubes to fill the hollow cube.

7. Erase digits so as to place as many zeroes as possible in the leftmost places. In this way, the total number of digits of the number will shrink:

10780952043600

The digits we erased have the sum: $1 + 7 + 8 + 9 + 5 = 1 + 9 + 7 + 8 + 5 = 10 + 15 + 5 = 30$.

Copyright Goods of the Mind, LLC, 2014

8. Convert Kangaroy's jumps into Kangarob-size jumps. 30 Kangaroy-size jumps are the same as 60 Kangarob-size jumps. Each minute, Kangaroy gets $60 - 40 = 20$ Kangarob-size jumps farther ahead of Kangarob. Multiply this by 3 to get the answer.

9. They have found 7 numbers: 17, 26, 35, 44, 53, 62, 71. Because both numbers that are added are 2-digit numbers, we have not counted the number 80 ($80 + 8 = 88$.)

10. If we cut the corners along a square the perimeter does not change.

11. The largest product is obtained if the number has the digits 3, 3, 2, 2, in any order. This is because the product increases if we use more small digits than few large digits. However, we also see that $3 \times 3 > 2 \times 2 \times 2$ and, therefore, it is not advantageous to regroup 2 3s into 3 2s.

    In the case of a number with a digit sum of 10, there is a remainder of 1 when we divide 10 by 3: $10 = 3 + 3 + 3 + 1$. When we multiply these digits we notice that a 3 and a 1 are not as good as 2 2s: $3 + 1 = 2 + 2$ but $3 \times 1 < 2 \times 2$.

12. A fourth a number goes 4 times into that number.

13. D is the obvious choice, since the circle in the second row cannot move from the side of the card to the center of the card.

14. The number of ones increases by one - therefore, the number of ones must be a triangular number. Since there are also zeroes, locate the triangular number closest to 27. 21 is such a number since : $1 + 2 + 3 + 4 + 5 + 6 = 21$. The corresponding number of zeroes would be 6. Check that this number matches the total number of digits: $21 + 6 = 27$ The digit sum is 21.

Copyright Goods of the Mind, LLC, 2014

15. The glass pellets weigh $550 - 250 = 300$ grams more than the milk. Half of the glass pellets weigh 150 grams more than half of the milk.

16. The 1st century CE spanned the first one humdred years from 1 to 100. The second century CE spanned from the year 101 to the year 200, etc. The 10th century CE started on January, 1st, 901.

17. There is a graduation date on each of the 7 years $(2012 - 2006 + 1)$. Each year, 100 students graduate. Therefore, 700 students graduated from 2006 to 2012, inclusive.

18. The numbers from 30 to 39 will be the same as the numbers from 70 to 79 Also, $43, 53, 63, 73 \, 83$, and 93 will be the same as their counterparts that end in 7. From the total of 100 numbers, only $100 - 10 - 6 = 84$ numbers will be different.

19. Make a drawing of the numbers placed around the circle. Since we do not have any values, let us make the numbers different colors. Place three different color 'numbers' on the circle. Since any three neighbors have the same sum, we notice that the next number is already known: it must be the number that drops out of the group of three when we add up the new neighborhood of three. Similarly, from here on, the numbers must repeat themselves like in the figure:

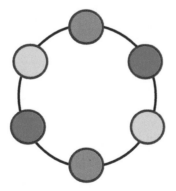

The smallest number of numbers we can use is 6 (since there are more than three.)

20. The row of letters is transformed as follows:

    1 ABCDDCBA

    2 CCDDDDCC

    3 CDDDDDDC

    4 DDDDDDDD

Daniel has performed 3 operations: 1-2, 2-3, 3-4.

21. For example, let the colors be red, green, blue, and purple. There are only 3 choices for the color opposite the red: green, blue, and purple. The positions of the other two colors do not matter, because they will be switched if we flip the bracelet. The bracelets in the figure are identical - flip one of them upside-down to get the other.

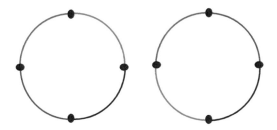

22. Make a diagram:

Copyright Goods of the Mind, LLC, 2014

Start

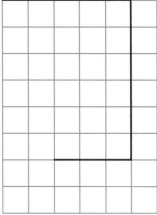

Kangaroy has to hop 2 miles West and 6 miles North in order to return. No matter how he chooses to turn, the shortest path to the starting point is $6 + 2 = 8$ miles long.

23. For the last digit of a number to remain the same when the number is multiplied by 9, the digit can only be 0 or 5. Since 0 is not acceptable, the digit must be 5.

24. They met in the middle, since they tunnelled at the same speed. Wor tunnelled through the whole first volume plus another 70 pages from volume two. Orm tunnelled through half of 886, which is 443 pages. This means the first volume must have $443 - 70 = 373$ pages.

Copyright Goods of the Mind, LLC, 2014

1. Add the digits of 12340 to find 10 as the sum of its digits. This means, the remainder of dividing 12340 by 3 is 1. Therefore, add 2 to find the next multiple of 3: 12342.

2. If any number is multiplied by an even number, the result is even. The only answer choice that is not even is C.

3. The hour hand makes a complete rotation in 12 hours. The minute hand makes a complete rotation in one hour.

4. When the seventh kangaroo jumps from last to first, the first kangaroo is again last and all the kangaroos are in the same order as at the start.

5. The neighbors differ by 2. Subtract 2 from 136 and divide by 2 to find the smaller neighbor: 67. The number is 68 and its digit sum is 14.

6. Write an alphabet and draw a line that divides it into two equal parts: from A to M and from N to Z. Each half consists of 13 letters. As they hop at the same time, they will cross the line at the same time and continue to hop on. Count the same number of letters on each side of the midline. When Kangarob reaches Q, Kangaroy reaches J.

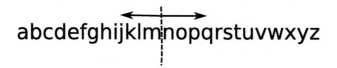

abcdefghijklmnopqrstuvwxyz

7. The smallest possible number of costume changes is 2.

8. For two students to face each other the total number of students has to be even. If the number of students is odd, there is no student diametrically opposite from another student. Note that this happens only if the students are equally spaced around the circle.

9. Any of the A-D choices can be obtained:

$$5 + 2 \times 4 = 13$$
$$5 \times 2 + 4 = 14$$
$$(5 + 2) \times 4 = 28$$
$$5 \times (2 + 4) = 30$$

We can also obtain:

$$(5 + 4) \times 2 = 18$$
$$5 \times 4 + 2 = 22$$

Copyright Goods of the Mind, LLC, 2014

10. Make a sequence with all the filled quantities. Each subsequent bottle contains 50 mL less than the previous one:

$$1000, 950, 900, 850, \ldots$$

In the sequence, the bottle with only 450 mL in it is the 12th one.

11. On the initial grid, only one unit square can be drawn:

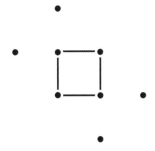

On the extended grid we can draw 3 more unit squares:

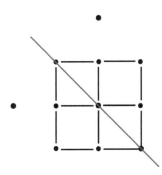

12. Let us say that the stars that are not white or red are yellow. Red and yellow stars add up to 13, while white and yellow stars add up to 9. The number of red stars is equal to the total minus the white and yellow: $20 - 9 = 11$. The number of white stars is: $20 - 13 = 7$. The number of yellow stars is $20 - 7 - 11 = 2$.

Copyright Goods of the Mind, LLC, 2014

13. For choice B, the triangle and the rectangle do not share a whole side:

Note that choice D is possible, if the triangle is completely inside the rectangle:

14. After the first operation, Julian has:

$$6 + 12 = 18 \text{ strings}$$

After the second operation, Julian has:

$$9 + 18 = 27 \text{ strings}$$

15. In decimal notation, the last two digits of a multiple of 4 form a number that is also a multiple of 4. Two-digit multiples of 4 that can be formed with the given digits are: $12, 16, 28,$ and 68. Each of these can form the last two digits of a number of interest, while the first two digits can be chosen anyhow from the remaining digits.

Copyright Goods of the Mind, LLC, 2014

Since there are only 2 digits remaining in each case, there are two possibilities for the first two digits for each of the combinations available as last digits, to a total of 8.

$$6812, 8612$$
$$2816, 8216$$
$$6128, 1628$$
$$1268, 2168$$

16. In figure (A), by cutting the green ring, the red and blue rings remain linked.
    In figure (B), by cutting the blue ring, the orange and green rings remain linked.
    In figure (C), by cutting the orange ring, the green and blue rings remain linked.
    Figure (D) fulfills the condition.
    In figure (E), by cutting the orange ring, the green and blue rings remain linked.

Copyright Goods of the Mind, LLC, 2014

17. Divide all the numbers by 3, as the problem will be simpler if we use smaller numbers:

1				
2	3			
4	5	6		
7	8	9	10	
etc.				

Row 1, Row 2, Row 3 label the first three rows.

and try to find the row for the number: $\dfrac{111}{3} = 37$.

Notice that, in the first column, the numbers follow a pattern:

$$1$$
$$2 = 1 + 1$$
$$4 = 2 + 2$$
$$7 = 4 + 3$$

in which the next rows are going to start with the numbers: $7 + 4 = 11$, $11 + 5 = 16$, $16 + 6 = 22$, $22 + 7 = 29$, $29 + 8 = 37$. It follows that 37 is the first number on the 9th row. This row number remains the same if the multiply all the numbers by 3.

18. Pairs of numbers have to have the same sum - this sum can be either even or odd. There are 4 even numbers in the list and 1 odd number. It is not possible to form 3 odd sums, therefore the sum must be even. The unknown number must be odd and, moreover, it must be paired with 9 to form an even sum.

Copyright Goods of the Mind, LLC, 2014

19. Five times the double of a number is the same as ten times the number. Adding this to the triple we get 13 times the number. Make a diagram to be sure your reasoning is correct:

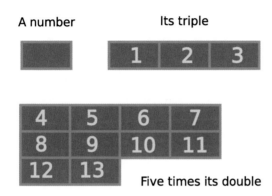

We have to find out which of the answer choices is a multiple of 13:

8 is not a multiple of 13

$32 = 13 \times 2 + 6$

$80 = 13 \times 6 + 2$

$91 = 13 \times 7$

$95 = 13 \times 7 + 4$

20. 99 numbers must be placed in groups of 6. Divide 99 by 6 and find the remainder 3. This means that some hexagons will be completely filled and the 3 last numbers will be left over. Since each new hexagon starts in the direction A, the last number - 1 - will be placed in the direction labeled C.

Copyright Goods of the Mind, LLC, 2014

21. There are 8 numbers in total, 6 of them will form 3 identical sums and 2 will be left over. What can the sums be? Add all the numbers to find out the total: $1 + 2 + 3 + 4 + 5 + 6 + 7 + 8 = 36$. The largest sum possible is $8 + 6 = 14$ (7 and 8 are on the same card.) But $14 \times 3 = 42$, which is larger than 36. We have to look for smaller sums. Use integer division:

$$
\begin{aligned}
36 &= 3 \times 11 + 3 \\
36 &= 3 \times 10 + 6 \\
36 &= 3 \times 9 + 9 \\
36 &= 3 \times 8 + 12 \\
36 &= 3 \times 7 + 15
\end{aligned}
$$

etc. The first does not work since 1 and 2 are on the same card. The third one does not work because, if 2 and 7 are paired, 8 and 1 cannot be paired because they are on the same cards as 2 and 7. From the fifth one on, the remainder is larger than 14, which is the largest possible sum of two numbers. We find that the second and the fourth cases work - the arrangements Kalynka made are:

$$[1, 2]\,[8, 7]\,[3, 4]\,[6, 5]$$

$$[8, 7]\,[1, 2]\,[6, 5]\,[3, 4]$$

22. First of all, note that the orientation of the letters CLG is not important since we cannot see them in the picture provided. We only have to make the letters BFF match. Now, consider the pairs of sides that will be glued together when the net is folded into a cube:

Copyright Goods of the Mind, LLC, 2014

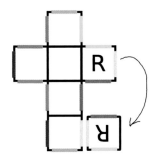

Especially look at how the two grey sides and the two dark blue sides correspond. An example is given with the letter R. Cut out the cell with the letter R and rotate it until the grey sides correspond. Look at what happens to the orientation of the letter.

23. If Nadja does not lose 3 bracelets, she has 16 less than Colleen. Since Colleen has 3 times more bracelets than Nadja, 16 must be twice the number of bracelets that Nadja has. Therefore, Nadja has 8 bracelets and Colleen has 24. Together, they have 32 bracelets.

24. The total number of students is $8 \times 4 + 4 = 36$. Subtract 3 for Tom, Huck, and Becky to find the number of students that separate them: $36 - 3 = 33$. Divide by 3 to obtain the number of students separating Tom from Becky: $33 \div 3 = 11$.

Copyright Goods of the Mind, LLC, 2014